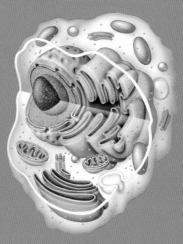

LIFE ON EARTH

CHELSEA HOUSE
P U B L I S H E R S
A Haights Cross Communications ◆ Company ®
Philadelphia

First hardcover library edition published
in the United States of America in
2006 by Chelsea House Publishers,
a subsidiary of Haights Cross Communications.
All rights reserved.

A Haights Cross Communications ✦ Company ®

www.chelseahouse.com

Library of Congress Cataloging-in-Publication
applied for.
ISBN 0-7910-9010-8

Project and realization
Parramón, Inc.

Texts
Eduardo Banquieri

Translator
Patrick Clark

Graphic Design and Typesetting
Toni Inglés Studio

Illustrations
Marcel Socías Studio

First edition - March 2005

Printed in Spain
© Parramón Ediciones, S.A. – 2005
Ronda de Sant Pere, 5, 4ª planta
08010 Barcelona (España)
Norma Editorial Group

www.parramon.com

TABLE OF CONTENTS

LOOKING FOR ANSWERS

This book offers an exciting look at the many forms of life on Earth.

The introduction to the book tries to answer questions such as "What is life?" and "How did life begin?" before continuing with a description of the cell as the basic unit of every living thing. The organization of the cell is described in detail, and we then look at how cells combine to form tissues, organs, and organisms. A simple classification of living things allows us to look at the many forms of life on Earth today. We have chosen to discuss only those living things closest to our daily lives—flowering plants and mammals.

Finally, we look at the basic laws of genetic inheritance, and at the changes living organisms go through over time, with a focus on the evolution of human beings.

Although we cannot discuss every aspect of life in these few pages, we feel that the themes we have chosen create a good basis for young readers who want an introduction to biology.

WHAT IS LIFE?

It is not surprising that life began in water, given that the depth of the seas provided protection against the harsh conditions on the surface of the Earth.

AN ENDURING MYSTERY

It is not easy to answer the question of what life is, since life is the most complex and mysterious phenomenon in nature. Living matter is formed by a series of inert chemical components. These components are organized in ways that lead to vital processes and characteristics, such as reproduction, growth, heredity, nutrition, sensitivity, metabolism, and the ability to adapt to an environment. These processes and characteristics distinguish animals and plants from nonliving things.

Of all these processes and characteristics that distinguish living organisms from nonliving matter, the ability to reproduce is perhaps the most notable, since only living organisms can reproduce and create other beings like themselves. The process of reproduction goes hand in hand with another characteristic of life, heredity. Through the process of heredity, distinct physical and mental traits are passed from one generation to the next.

Nevertheless, reproduction alone cannot guarantee that an organism will survive, since environmental conditions change over time. To ensure its continued survival, an organism must be able to adapt to its environment.

Sensitivity to stimulation is another characteristic that is unique to living creatures. Every living thing is able to respond in some way to stimuli in the world around it.

Finally, another important characteristic of living matter is its ability to renew and maintain its chemical makeup. This property is called metabolism.

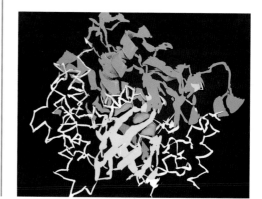

Proteins were the first complex molecules that developed in the depths of the primitive seas; protozoans developed from these proteins.

The oldest known fossil remains of Earth's earliest life-forms are of *Cyanophycea* (blue algae), dating back more than 2.4 billion years. It is believed that Earth's first life-forms appeared long before that.

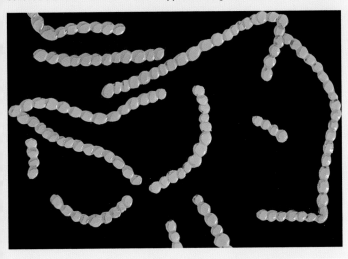

Reproduction is one of the characteristics that separate nonliving things from living things.

HOW DID LIFE APPEAR ON EARTH?

This is a difficult question to answer, as we don't have any fossil remains of the very first living things. Not only do we lack fossils of the type of structures that came before the first cells, but we don't even have access to sediments (rock remains) that would allow us to reconstruct the kind of environment that existed on the primitive Earth. The question of how life began has long been considered one of the most important questions in science. Over the course of history, various hypotheses have been proposed to explain how life began. One of the first explanations used the idea of "spontaneous generation," in which nonliving things were believed to produce living things.

Today, the most widely accepted hypothesis is that life began in the water, since the depth of the seas provided protection against the harsh conditions of the Earth's surface. The disappearance of oxygen from the atmosphere was very important for the development of life, because it allowed a new atmosphere to form. This new atmosphere was rich in water vapor, ammonium, methane, hydrogen, nitrogen, and other gases that created the conditions necessary for the formation of organic molecules.

Russian biochemist Alexander Oparin suggested in 1924 that organic molecules may have developed over time to become more and more complex, and that these new, more complex organic molecules changed over time. According to this theory, the oceans at first contained a large number of dissolved organic compounds. During a long process, these molecules combined to form larger molecules, and these, in turn, formed temporary complex molecules such as amino acids. Some of these complex molecules formed a proto-cell (the precursor of the first living beings). The proto-cell then acquired certain traits, including the ability to isolate itself from the surrounding environment, to bring certain surrounding molecules inside itself, and to release others. According to this view, metabolic functions, reproduction, and growth began after the proto-cell gained the ability to take in molecules and make them part of its structure. Ultimately, the proto-cell was able to separate parts of itself into multiple new units of life, each with its own characteristics—in other words, the proto-cell was able to reproduce.

Since the first living creatures lived in the sea and did not have hard skeletons, their remains have not been preserved as fossils.

The great mass extinction at the end of the Cretaceous Period may have been caused by the crash of a large meteorite.

THE BASIC UNIT OF LIFE

The cell is the smallest independently functioning unit of an organism. All living organisms are made up of cells, and it is generally agreed that no living thing consists of less than a single cell. Some microscopic organisms, such as bacteria and protozoans, are single-celled, while animals and plants are made up of many millions of cells organized as tissues and organs.

Despite many differences in form and function, all cells are wrapped in a membrane (called a plasma membrane) that encloses a water-rich substance known as cytoplasm. Inside cells, many chemical reactions take place. These reactions allow cells to grow, produce energy, and eliminate waste. Together, all of these reactions are called metabolism. All cells contain hereditary information encoded in molecules of deoxyribonucleic acid (DNA). This information directs the activity of the cell and ensures that traits will pass through reproduction from one generation to the next.

Prokaryotic and eukaryotic cells differ in both size and internal organization. Prokaryotes, a group that includes bacteria and cyanobacteria, are small cells, between 1 and 5 µm (micrometers; a micrometer is one one-thousandth of a millimeter) in diameter, with a simple structure. Eukaryotic cells, which form all other living organisms, including protozoans, plants, fungi, and animals, are much bigger (between 10 and 50 µm long), and have their genetic material wrapped in a membrane that forms a spherical organelle called a nucleus.

CLASSIFICATION OF LIVING THINGS

The Swedish naturalist Carolus Linnaeus (1707–1778) invented the classification system we use today. This system places living things in categories with increasing degrees of precision. That is, it puts them in categories that gradually become more restrictive until, at last, we get to a category in which there is only one type of organism.

To study the way living things were classified, a new science arose, known as taxonomy.

Life has diversified, so that we find living things in even the most hostile environments on our planet.

Pine trees are the most common and well-known conifers. Pine trees grow in great abundance in a variety of soil types and climates.

Taxonomy groups living things according to their similarities and differences when compared with other living things. These units are organized from large to small in seven hierarchical categories, which are Kingdom, Phylum (Type), Class, Order, Family, Genera, and Species.

These seven levels are sometimes too broad to classify all living things according to form. In some cases, scientists have to create intermediate subdivisions, such as superorder, suborder, and superfamily.

The basic unit from which all classification begins is the species, which is defined as a biological unit for grouping organisms. Within a species, all members are able to mate with each other and create viable and fertile offspring. To designate a species, we use a name made up of two words, written in italic letters. The first of these words, which starts with a capital letter, corresponds to the genera, and the second, which begins with a lowercase letter, refers to the species. Sometimes it is necessary to define a subspecies. In this case, we need to add a third word, also written in lowercase letters—for example, *Homo sapiens neanderthalensis*, in which *Homo* is the genera, *sapiens* is the species, and *neanderthalensis* is the subspecies.

As an example, we show the classification of the domestic dog in the box below. Despite the differences in appearance among existing kinds of dogs, all dogs are part of a single species called *Canis familiaris*, since it is theoretically possible for any two dogs to mate, even a chihuahua and a mastiff. The offspring of that match should also be able to breed.

Superkingdom	Eukaryota
Kingdom	Animalia
Phylum (Type)	Chordata
Subtype	Vertebrata
Class	Mammalia
Order	Carnivora
Family	Canidae
Genera	*Canis*
Species	*Canis familiaris*

FROM INERT MATTER TO LIFE

Life is believed to have begun 3.8 billion years ago, with life-forms similar to the simplest forms known today, such as viruses, bacteria, and blue-green algae. The atmosphere at the time of life's origin lacked oxygen, and was made of methane, ammonium, hydrogen sulfide, water vapor, hydrogen, and helium. Chemical compounds of these gases gave rise to the first organic molecules, which are the basis of every living thing. The transformation from organic material to living things took place in the sea, in ways that are still not fully understood.

■ reducing atmosphere

■ ultraviolet rays

■ organic synthesis
the chemical combination of the gases of the primitive atmosphere formed the first organic molecules

■ primitive soup
the temperature of the rich "soup" of organic material in the waters of primitive Earth favored reactions among molecules, which joined to form compounds of greater size and complexity

■ diversification
simple organic molecules excited by solar radiation and electrical discharges continued to diversify and accumulated in the sea ("primitive soup")

■ the first organisms
were anaerobic (there was no oxygen in the atmosphere) and heterotrophs (they fed on the abundant organic material that existed in the primitive sea)

oxygen-rich ■ atmosphere

photosynthesis ■
the first photosynthetic algae appeared; the algae gave off oxygen, causing a change in the primitive atmosphere

reproduction ■
nucleic acids, and with them, the mechanisms for reproduction appear

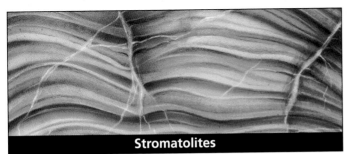
Stromatolites

Stromatolites are rolled-up sedimentary structures that are the product of metabolic activity by microorganisms (mainly bacteria or algae) whose presence has been identified in rocks up to 3.5 billion years old.

THE IMPORTANCE OF A REDUCING ATMOSPHERE

Unlike Earth's current oxygen-rich atmosphere, Earth's early atmosphere likely contained little oxygen and was a reducing atmosphere. This reducing atmosphere contained gases that gave rise to the first organic molecules that were essential for life to form on Earth.

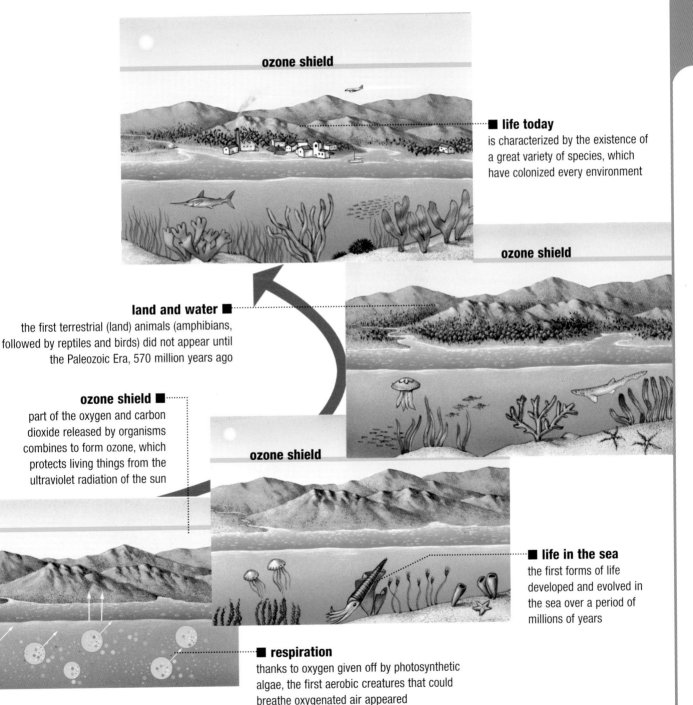

ozone shield

■ life today
is characterized by the existence of a great variety of species, which have colonized every environment

ozone shield

land and water ■
the first terrestrial (land) animals (amphibians, followed by reptiles and birds) did not appear until the Paleozoic Era, 570 million years ago

ozone shield ■
part of the oxygen and carbon dioxide released by organisms combines to form ozone, which protects living things from the ultraviolet radiation of the sun

ozone shield

■ life in the sea
the first forms of life developed and evolved in the sea over a period of millions of years

■ respiration
thanks to oxygen given off by photosynthetic algae, the first aerobic creatures that could breathe oxygenated air appeared

FROM ELEMENTARY PARTICLES TO COMPLEX ORGANISMS

The matter that forms living things is organized on different levels. We can only truly say a thing is "alive" once it has reached a certain degree of complexity. Even so, every substance that makes up part of an organism has a very special function. In the system of organization used for living things, each level includes all the levels that came before it. Living things are described in a synthetic manner, which means they pass from the simple to the complex.

BIOLOGICAL ELEMENTS

This term is used to describe the "molecular building blocks" of living matter, because with them, it is possible to build biological molecules, the basic units that make up all living creatures. The most common biological elements are carbon (C), nitrogen (N), hydrogen (H), oxygen (O), sulfur (S), and phosphorus (P).

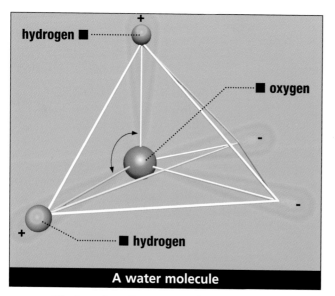

hydrogen ■

oxygen

hydrogen

A water molecule

Life depends on the presence of water, and water makes up most living things.

organism ■
human beings and all organisms that have many cells are formed by an integrated set of organs coordinated by a nervous system and a hormonal system

supramolecular association ■
molecules combine to form more complex structures such as membranes; the combination of these structures make up organelles

organelle ■
the combination of supramolecular structures (made of molecules) create organelles such as mitochondria, which are parts of a cell

■ organ
the gathering of tissues gives rise to specialized structures that perform certain functions, such as the skin, the stomach, the liver, and the heart

tissue ■
is formed by the gathering of cells that have specialized functions, such as the skin

molecule ■
is formed by the combination of two or more atoms of the same or different types; examples include proteins, glucose, and water

atom ■
is the basic unit of matter; it is formed by three types of particles: protons, neutrons, and electrons

cell ■
is the basic unit of living things; groups of cells form tissues

THE SIMPLEST FORMS OF LIFE

Prokaryotic cells do not have a real cell nucleus. They have no nuclear membrane to separate their genetic material from their cytoplasm. They are always unicellular (one-celled) creatures. Prokaryotic cells were the first cell type to appear on Earth (3.8 billion years ago). Their structure is characteristic of simple organisms such as bacteria. Even simpler organisms are viruses. Viruses do not have their own metabolism. Instead, they rely on the metabolism of a host to reproduce. For this reason, viruses fall into a category between living and nonliving things.

mesosomes ■
rough parts in the plasma membrane that increase its surface area; this is where cellular respiration takes place

ribosomes ■
are organelles that make the cell's proteins

cytoplasm ■
is a thick, semi-liquid substance where the ribosomes are found

nucleoid ■
is the area that holds the chromosome; it lacks a cell membrane, and therefore, looks like it is spread out within the cell

capsule ■
many bacteria have a thick third covering that looks like gel and is made up of sugars

bacterial wall ■
defends bacteria from the action of antibiotics

SUCCESSFUL CREATURES

Bacteria are one of the groups that has achieved the greatest degree of biological success, both because they are one of the most common life-forms and because they adapt to many kinds of environments and ways of life. To give you an idea of how widespread bacteria are, consider that a single gram of soil may contain 25,000 bacteria.

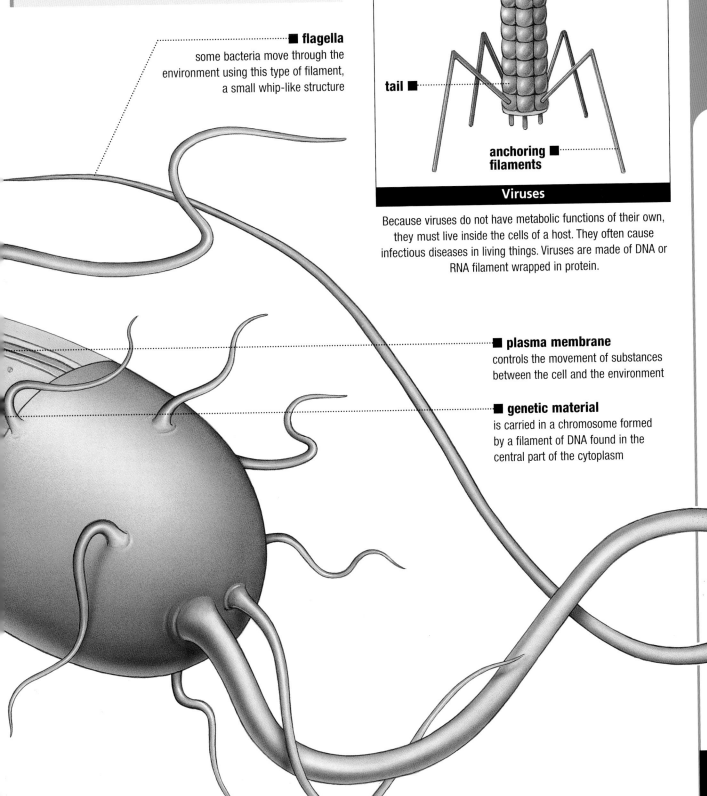

flagella
some bacteria move through the environment using this type of filament, a small whip-like structure

head ■

■ neck

tail ■

anchoring ■ filaments

Viruses

Because viruses do not have metabolic functions of their own, they must live inside the cells of a host. They often cause infectious diseases in living things. Viruses are made of DNA or RNA filament wrapped in protein.

■ plasma membrane
controls the movement of substances between the cell and the environment

■ genetic material
is carried in a chromosome formed by a filament of DNA found in the central part of the cytoplasm

THE BASIC UNIT OF LIFE

The structure of eukaryotic cells is much more complex than that of prokaryotic cells. Eukaryotic cells are bigger and contain many specialized organelles to carry out different processes. Furthermore, their genetic material is held inside a membrane that makes up the cell nucleus. Eukaryotic cells have three main parts: the cell membrane, the cytoplasm, and the nucleus. Eukaryotic cells are of two types, depending on how they obtain nourishment: autotrophic or heterotrophic. Plant cells obtain nourishment through the process of photosynthesis and, therefore, are autotrophic. Animal cells obtain nourishment from organic material and, therefore, are heterotrophic.

PLANT CELL

liposomes ■
are pouches of digestive enzymes formed by the Golgi apparatus

nucleus
is the guiding organ for the cell's life, since it carries the instructions the cell needs to make proteins

Golgi apparatus ■
transforms the products stored in the endoplasmic reticulum and holds them outside

cytosol ■
has no structure and makes up the liquid part of the cytoplasm

vacuoles ■
act as a storehouse for waste products or excretions

chloroplasts ■
are the organelles where plant cells carry out photosynthesis

endoplasmic reticulum ■
is where proteins formed by the ribosomes are stored, and where they go through chemical changes that the cell needs to survive

plasma membrane ■
materials that are changed during cellular metabolism enter and leave through the cell membrane

cytoplasm ■
is all of the cellular material surrounded by the membrane, except for the nucleus; the organelles are found in this watery medium

mitochondria ■
hold the enzymes needed for cellular respiration

ribosomes ■
are the organelles where proteins are made according to the instructions that come from the nucleus

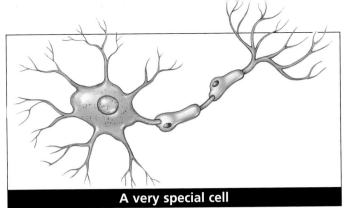

A very special cell

The neuron is the cell that makes up the basic functional unit of the nervous system, forming a tissue that sends messages about stimuli between different parts of an organism.

AUTOTROPHIC AND HETEROTROPHIC CELLS

Plant cells are autotrophs, which means that they are able to make their own organic material (food) by means of photosynthesis. Animal cells are heterotrophs, which means that they feed on organic material.

ANIMAL CELL

ribosomes ■
are organelles where proteins are made according to instructions that come from the nucleus

endoplasmic reticulum ■
is where proteins formed by the ribosomes are stored and where they go through chemical changes the cell needs to survive

Golgi apparatus ■
transforms the products stored in the endoplasmic reticulum and holds them outside

mitochondria ■
holds the enzymes needed for cellular respiration

plasma membrane ■
materials that are changed during cellular metabolism enter and leave through the cell membrane

cytoplasm ■
is all of the cellular material surrounded by the membrane, except for the nucleus; the organelles are found in this watery medium

cytosol ■
has no structure and makes up the liquid part of the cytoplasm

nucleus ■
is the guiding organ for the cell's life, since it carries the instructions needed for the cell to make proteins

liposomes ■
are pouches of digestive enzymes formed by the Golgi apparatus

centrosome ■
is the guiding organ for cells to sense the environment and move; plant cells do not have a centrosome

vacuoles ■
are storage areas for substances not used by the cell, although they usually have some other vital function, such as digestion or pumping out excess water

nucleole ■
nuclear structure that plays a role in forming ribosomes

THE TREE OF LIFE

Since the simplest forms of life came into existence some 3.8 billion years ago, living things have been going through changes from generation to generation in a process of evolution. Today, around 1.5 million different living species have been described, and more than 3 million probably exist. If we thought about the number of species that have evolved and are now extinct, the total number would be huge.

plant kingdom ■

these are multicellular (many-celled) eukaryotic organisms with tissues that have different functions; this kingdom includes mosses, ferns, flowering and nonflowering plants, and trees

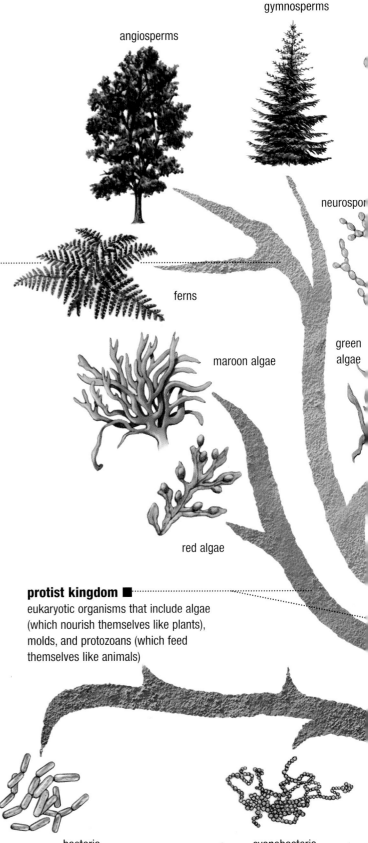

gymnosperms

angiosperms

neurospor

ferns

green algae

maroon algae

red algae

protist kingdom ■
eukaryotic organisms that include algae (which nourish themselves like plants), molds, and protozoans (which feed themselves like animals)

bacteria

cyanobacteria

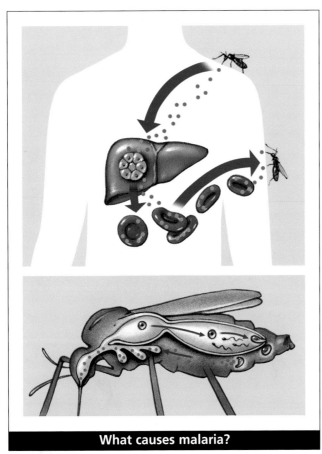

What causes malaria?

Malaria is an infectious disease caused by parasitic protozoans of the genera *Plasmodium*. This disease is spread to human beings by the bite of a female mosquito of the genera *Anopheles*.

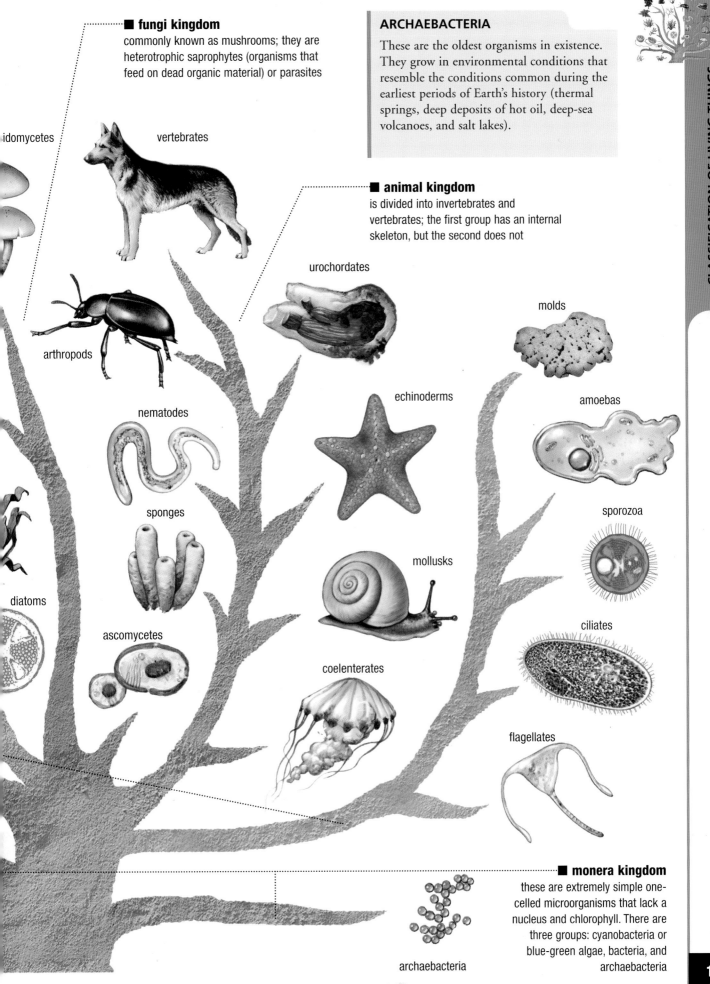

■ fungi kingdom
commonly known as mushrooms; they are heterotrophic saprophytes (organisms that feed on dead organic material) or parasites

ARCHAEBACTERIA
These are the oldest organisms in existence. They grow in environmental conditions that resemble the conditions common during the earliest periods of Earth's history (thermal springs, deep deposits of hot oil, deep-sea volcanoes, and salt lakes).

idomycetes

vertebrates

■ animal kingdom
is divided into invertebrates and vertebrates; the first group has an internal skeleton, but the second does not

urochordates

molds

arthropods

echinoderms

amoebas

nematodes

sporozoa

sponges

mollusks

diatoms

ciliates

ascomycetes

coelenterates

flagellates

■ monera kingdom
these are extremely simple one-celled microorganisms that lack a nucleus and chlorophyll. There are three groups: cyanobacteria or blue-green algae, bacteria, and archaebacteria

archaebacteria

A DIET OF MINERALS, WATER, AND SUNLIGHT

Relatively simple plants—fungi and algae, for example—lack the tissues and organs found in more complex plants with roots, stems, and leaves. More complex plants need roots, stems, and leaves to protect them from drying out, to distribute food, and to make their plant bodies stronger.

■ **leaves**
are the organs that help the plant breathe and carry out photosynthesis

flowers ■
contain the reproductive organs of the plant

stem ■
holds up all the organs of a plant (leaves, flowers, and fruits) and sends sap from the roots to the leaves and flowers

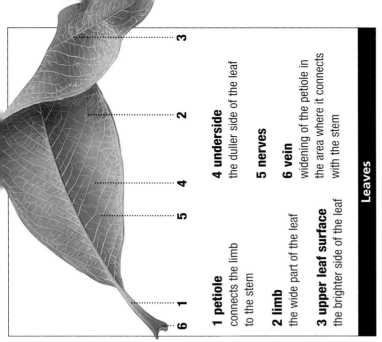

3

2

4

4 underside
the duller side of the leaf

5

5 nerves

6

6 vein
widening of the petiole in
the area where it connects
with the stem

1

1 petiole
connects the limb
to the stem

2 limb
the wide part of the leaf

3 upper leaf surface
the brighter side of the leaf

Leaves

They play a vital role in plant nutrition, since these are the organs that perform photosynthesis by taking in carbon dioxide and giving off oxygen.

MUSHROOMS

Mushrooms are a special type of simple plant. Because they lack chlorophyll and cannot perform photosynthesis, they must live off organic matter that is decomposing (breaking down).

fruit ■
is the part of the
plant whose job is to
protect the seeds
and make sure they
are spread around

■ **seeds**
are mature ovules,
from which, given the
proper conditions, new
plants will grow

root ■
its function is to absorb
substances that are
supposed to nourish the
plant (water and mineral
salts) and carry them up
to the stem

EARTH'S MOST DOMINANT ANIMALS

The animal kingdom includes all multicellular organisms that get energy by digesting nutrients and contain cells that are organized into tissues. The animals that have become the most biologically successful are the mammals. This success is due to the development of a viviparous, or live birth, system—that is, the development of the embryo inside the mother. This enables the nervous system to be highly developed by the time the offspring is born. The most successful mammals are those of the primate family. Humans are part of the primate family and occupy the highest level in the biosphere.

spinal column ■
is the main support of the body and covers and protects the spinal cord

kidney ■
its main function is to purify the blood and produce urine

The only animals that drink milk

A distinguishing feature of mammals is the ability of female mammals to nourish their young independently with milk produced in their mammary glands. Milk gives energy to the newborn and helps it grow.

bladder ■
urine, produced by the kidneys, is carried to the bladder, where it is stored until it is excreted

anus ■
is the end of the digestive tube, where waste is excreted

reproductive organs ■
organs where animals' germ (sex) cells mature and are stored

THE FUR OF MAMMALS

Fur allows mammals to live in nearly all climates, since it protects them from the cold, keeps body heat from escaping, and keeps them safe from wind, rain, and solar radiation. Many mammals have colored skin that helps them blend in with their environment (an ability known as mimicry) and hide from their prey and their enemies.

■ **liver**
processes blood coming from the stomach and intestines, and breaks up nutrients into forms that are easier for the rest of the body to use

■ **brain**
controls the central nervous system

■ **nasal cavity**
the inside of the nose may be divided by the septum into two cavities through which air enters during breathing

■ **intestine**
extends from the stomach to the anus and serves mainly to complete digestion and prepare for the elimination of waste

■ **mouth**
is the first part of the digestive tube, through which food is taken into the body

■ **pharynx**
is the continuation of the mouth. It connects to the nose through two holes, with the ear through two other passageways (the Eustachian tubes), and with the respiratory tube (trachea). When food passes through, the trachea shuts by means of a valve called the epiglottis, which prevents the food from blocking the respiratory tube

■ **esophagus**
is the part of the digestive tube through which foods pass from the pharynx to the stomach

stomach ■
here, food is turned into substances that can be dissolved and taken in by the tissues

■ **lungs**
are two organs located in the thoracic cavity that make breathing possible

■ **heart**
pumps the blood through the bodies of animals and regulates blood circulation

THE CREATION OF A NEW LIFE

Sexual reproduction is the most common method of reproduction in multicellular animals. This type of reproduction is carried out through special cells called "germ cells" or gametes. These cells are formed in the sex organs. Their function is to merge to form a zygote, which develops into offspring that has a combination of the traits of both parents.

■ ovary
is the organ where eggs (the female gametes) are produced and released through a process known as ovulation

■ egg
female gametes develop into an embryo after fertilization

■ uterus
the fertilized egg arrives at the uterus, where it is implanted and begins to develop. This development continues until the baby is ready to be born

■ fallopian tubes
after an egg cell has been released from the ovary, it enters one of the fallopian tubes, where it is fertilized by a sperm cell

sperm cells ■
male gametes whose aim is to fertilize an egg cell

vagina ■
female genital organ, through which the male introduces sperm cells

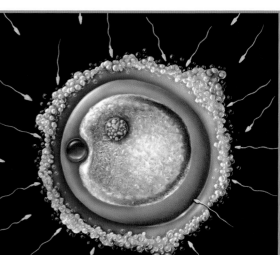

A RACE WITH A SINGLE WINNER

The fertilization of an egg by a sperm cell is a race with a single winner. In human ejaculation, more than 300 million sperm cells are deposited. Between 50 and 150 make it to the egg in the fallopian tube, and only one of these is able to fertilize the egg.

a difficult road ■
sperm cells have flagella (long, thin outgrowths of the cell) that enable them to move and make contact with the egg

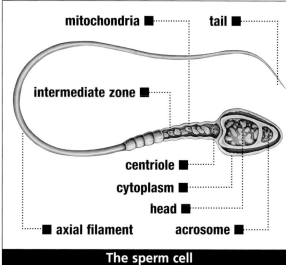

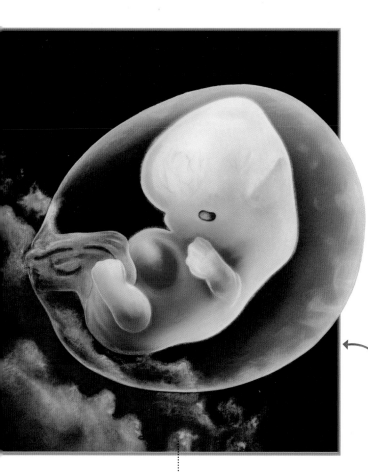

mitochondria ■ ⋯⋯⋯⋯ **tail** ■

intermediate zone ■ ⋯⋯

centriole ■ ⋯⋯⋯

cytoplasm ■ ⋯

head ■ ⋯

■ **axial filament** **acrosome** ■ ⋯

The sperm cell

The top part of a sperm cell, called the acrosome, allows it to penetrate an egg. The middle part of the sperm cell contains the mitochondria, which supply the energy that the sperm cell needs to move.

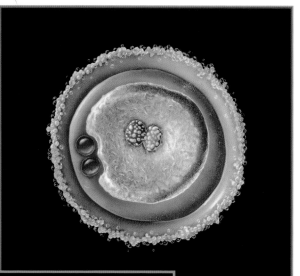

embryo ■ ⋯⋯⋯⋯
the zygote begins to divide, and the new cells that are formed differentiate into tissues, organs, and systems to create a new individual

zygote ■ ⋯⋯⋯⋯
the nuclei of the sperm cell and the egg combine, thus completing the process of fertilization

the winner ■ ⋯⋯⋯⋯
only one sperm cell manages to penetrate and fertilize the egg

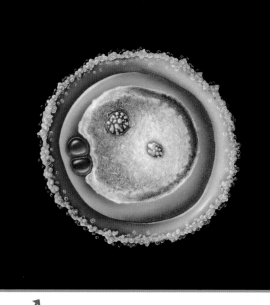

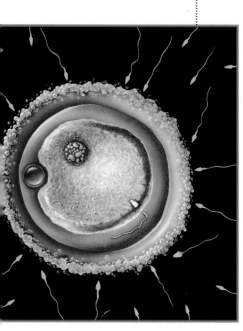

■ **egg cell**
once inside the egg, the sperm loses its tail and its nucleus gets larger

THE WONDER OF FLOWERS

The flower is the reproductive organ of certain plants. It gives way to fruits, which in turn hold seeds; however, not all plants that form seeds have flowers. Flowers help a plant get pollinated and protect its seeds until they are detached. Flowers are fertilized when a grain of pollen, which contains the male gametes, arrives at the stigma and descends by means of the style to reach the ovule (female gamete). This process produces the seed from which a new plant will grow.

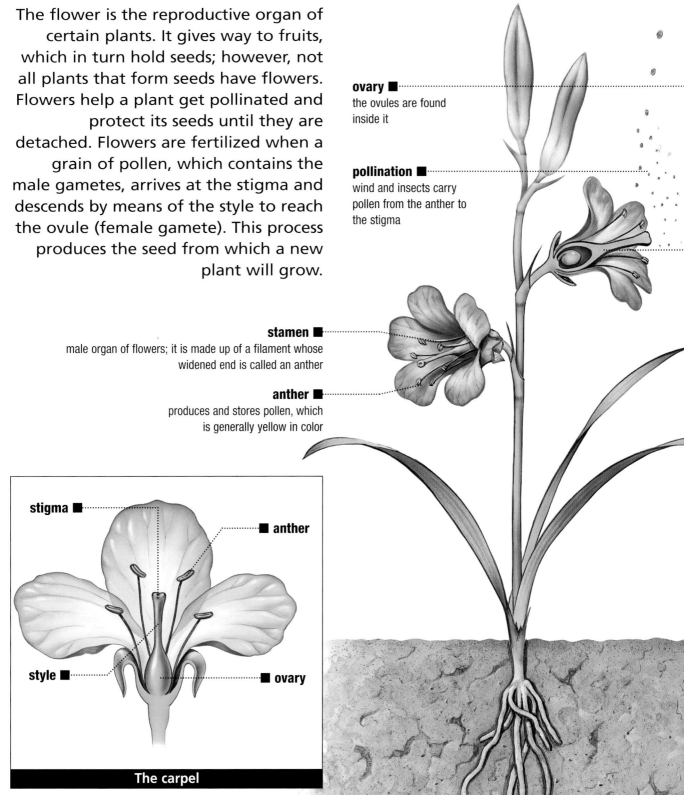

ovary ■
the ovules are found inside it

pollination ■
wind and insects carry pollen from the anther to the stigma

stamen ■
male organ of flowers; it is made up of a filament whose widened end is called an anther

anther ■
produces and stores pollen, which is generally yellow in color

stigma ■

■ anther

style ■

■ ovary

The carpel

The female reproductive part of the flower is a leaf that has been modified and that still has its green color. In a single flower, there may be one or more carpels; the set of these is called the pistil.

stigma
is the sticky surface at the top of the style; it traps and holds pollen swept out by the wind or by insects

style
is a tube-like structure that holds the stigma; the style leads below, where the ovary containing the ovules is found

NOT ALL PLANTS HAVE FLOWERS
Plants like pine trees do not have flowers and carry their ovules on the scales of a cone. These ovules are more exposed than those carried in the closed cavities of flowering plants.

pollen
tiny grains that contain sperm-like cells. They are produced by the male reproductive apparatus of flowers and carry the spermatic cells to the female reproductive structures to fertilize them

pistil
holds the female sexual apparatus, which has three distinct parts: stigma, style, and ovary

ovules
after the ovules are fertilized by pollen, seeds sprout; these seeds will form a new plant

fertilization
a grain of pollen arrives at the stigma and travels down the style until it reaches the ovule; then the zygote is formed. The maturation of the zygote will give origin to the seed

fruit
organ resulting from the development of the ovary in flowering plants; it is formed by the seeds and protective coverings

seeds
are mature ovules, from which new plants will arise under the right conditions

germination
starting with a seed, a set of events takes place, leading to additional growth or development of the plant

WHY DO WE LOOK LIKE OUR PARENTS?

In 1865, Gregor Mendel, an Augustinian monk from Austria, came up with the laws of inheritance that bear his name. These were the result of studies he began after a discovery he made in his garden with peas. Mendel called the agents of biological inheritance "factors." Today, these "factors" are known as genes, and any hereditary trait is determined by two genes, one from the father and the other from the mother.

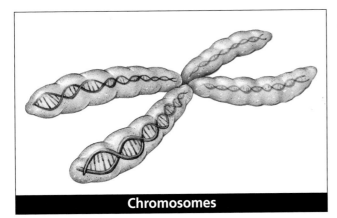

Chromosomes

These are cellular structures formed by DNA and proteins, which are charged with transmitting hereditary traits from one cell to another. They are made up of a set of genes that present themselves in pairs (homologues).

DNA TESTING

The genetic material of any individual, living or dead, is made of DNA. DNA can be used to determine who someone's father is or the identity of a corpse because every individual inherits the DNA of his or her parents—exactly half from each parent.

P ■
initial paternal generation

homozygote ■
individual that, for a given gene, has the same type of allele in each homologous chromosome; for example, AA or aa

allele ■
the gene that controls the color of the pea seed has two alleles, one that determines yellow color (AA) and another that determines green color (aa)

F1 ■
first generation

heterozygote ■
individual that, for a given gene, has a different allele in each homologous chromosome; for example, Aa

dominant trait ■
is the trait that is manifested. In this case, the color yellow is dominant over green

Mendel's first law

When there is a cross between individuals of the same species that belong to pure races (homozygotes), all the hybrids of the first generation are the same.

■ **phenotypes**
are traits directly observable by our senses that arise due to interactions between the genotype and the environment

■ **genotype**
is the set of genes that an organism inherits from its parents. Half of its genes come from the father, and the other half come from the mother

■ **recessive trait**
is a trait that is not expressed and remains hidden. This seed is a carrier of the allele for green, but does not show it, although it will transmit this allele to the next generation

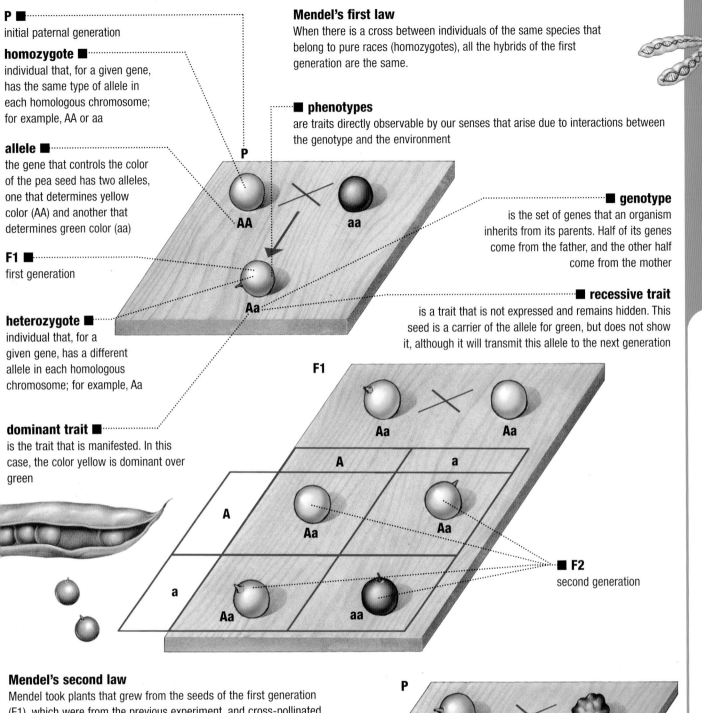

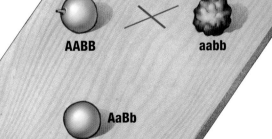

Mendel's second law

Mendel took plants that grew from the seeds of the first generation (F1), which were from the previous experiment, and cross-pollinated them with each other. From this cross, he got yellow and green seeds in the numbers shown in the illustration. According to his results, although the allele that determines green color in the seeds seemed to have disappeared in the first generation, it returned to express itself in this second generation.

Mendel's third law

In cases where two different traits are examined, each of them is passed on, according to the previous laws and independently of the presence of the other trait. To demonstrate this, Mendel crossed smooth and yellow peas with green and wrinkled peas (both homozygotes for the two traits). In accordance with the first law for each of the traits under consideration, the peas he got from this cross were all yellow and smooth, which revealed that the dominant alleles for these traits are those that determine yellow color and a smooth shape.

WHERE DO WE COME FROM?

The first living things were very simple, but over time, their descendants became more organized and complex. This process is known as evolution. The human being of today is the result of an evolutionary process that began 5 million years ago, although *Homo sapiens* (today's humans) did not appear until 100,000 years ago.

Homo neanderthalensis ■
arose 150,000 years ago, and suddenly became extinct 30,000 years ago; had a large cranial capacity of 1,600 cubic centimeters and a compact and robust body structure

Homo sapiens ■
is the present-day human species; the oldest remains that have been found are about 100,000 years old, show a cranial capacity of 1,350 cubic centimeters, and have a great range of heights and weights

Ardipithecus ramidus ■
is the oldest hominid fossil (4.4 million years old); considered the first species separated from the chimpanzee lineage and our missing link

Homo erectus ■
had a cranial capacity of between 800 and 1,300 cubic centimeters; lived in Asia until less than 100,000 years ago, although may have been in existence as long as 1.6 million years ago

■ **Homo heidelbergensis**
arose some 500,000 years ago, and continued to live until 30,000 years ago; cranial capacity of this species was about 1,125 cubic centimeters; individuals of this species were large and were able to control fire

■ **Homo antecesor**
lived more than 800,000 years ago; was tall and strong, with a cranial capacity of about 1,000 cubic centimeters, and lacked a chin

THE ENIGMA OF THE NEANDERTHAL

How and why did Neanderthals suddenly disappear 30,000 years ago? There are two scientific concepts that attempt to explain this: Some scientists believe that Neanderthals evolved into *Homo sapiens*, while others believe they died out from natural causes because they were unable to adapt to new climatic conditions on the planet.

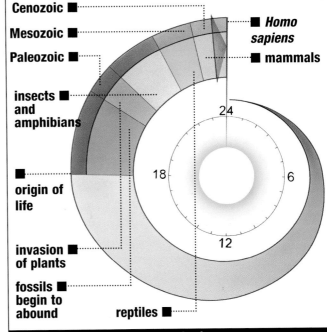

Cenozoic ■
Mesozoic ■
Paleozoic ■
insects ■ and amphibians
■ origin of life
■ invasion of plants
fossils ■ begin to abound
reptiles ■
■ *Homo sapiens*
■ mammals
24
18
6
12

Late arrivals

If we compress the history of the Earth into a single day (24 hours), we see that the first living things appear at 6:00 P.M., and human beings at 11:59 P.M. Despite this late arrival, human beings already dominate the Earth.

Australopithecus ■

lived between 4.2 and 1.2 million years ago; was bipedal (walked on two legs), small in stature (between 3 feet, 3 inches and 4 feet, 7 inches) and light in weight (40 to 65 pounds)

■ ***Homo habilis***

is the representative of the genus *Homo* and lived between 1.9 and 1.4 million years ago; had a cranial capacity of 650 cubic centimeters and made very simple stone tools

■ ***Homo rudolfensis***

had a brain of some 750 cubic centimeters in size and a face and jaw larger than those of *Homo habilis*; would have lived between 1.9 and 1.6 million years ago

■ ***Homo ergaster***

lived between 1.8 and 1.5 million years ago, and represents a great qualitative leap due to its large brain (850 cubic centimeters) and its tall stature (taller than 5 feet, 6 inches)

GLOSSARY

Chromosome	Filament-shaped corpuscle that is part of the cell nucleus and carries hereditary and genetic material, which it preserves and transmits.
DNA	Deoxyribonucleic acid. Controls cellular metabolism, determines the structure of proteins in the cell. DNA is needed for cell division, since the genetic information resides precisely in the DNA sequences.
Enzymes	Group of proteins produced by living cells that measure and induce the chemical processes of life without being altered.
Gametes	Male or female sexual cells that unite with each other during fertilization to form the zygote.
Gene	Hereditary unit that controls each trait in living beings. From a molecular point of view, a gene corresponds to a section of DNA that contains information for making a kind of protein chain.
Inert	Nonliving.
Meiosis	Cellular division in which the copying of chromosomes is followed by the division of two nuclei. Each one of the four resulting gametes receives half of the number of chromosomes (haploid number) of the original cell.
Metabolism	Set of chemical reactions that take place inside a cell and are translated into a constant renewal of living matter.
Multicellular	Formed by many cells.
Photosynthesis	Metabolic process carried out by certain cells in autotrophic organisms, by which organic substances are made from other inorganic substances using light energy. During photosynthesis, plants change carbon dioxide into oxygen.
Proteins	Substances made of amino acids that form part of the basic material of cells and of animal and plant material.
RNA	Ribonucleic acid. Part of the DNA is copied to RNA. RNA acts as a messenger to bring information to the cytoplasm, where ribosomes turn the genes into proteins. For this reason, RNA that can bring a message from the nucleus to the cytoplasm is called messenger RNA.
Sedimentary	Formed by or from the matter that settles at the bottom of liquid.
Zygote	Cell that results from the union of a sperm cell and an ovule or egg cell.

THE HISTORY OF LIFE

YEARS AGO . . .	EVENTS
4.6 billion–570 million (Precambrian)	The Earth is formed and gradually cools; its atmosphere lacks oxygen. The first bacteria appear 3.8 billion years ago. Blue-green algae develop and give rise to an atmosphere rich in oxygen. Protists (single-celled organisms) develop. At the end of this period, nonflowering marine plants and the first invertebrates appear.
570–510 million (Cambrian)	Invertebrates spread through all the oceans. Trilobites abound, and the first mollusks and brachiopods appear.
510–439 million (Ordovician)	The first crustaceans evolve, and fish-like invertebrates that lack fins and jaws appear.
439–409 million (Silurian)	The first fish with jaws appear, and coral reefs abound in the seas. The first plants appear on dry land.
409–360 million (Devonian)	Fish dominate life in the seas. The first insects and the first amphibians appear. The first terrestrial flower appears.
360–290 million (Carboniferous)	The climate is warm and humid, so the Earth is covered with large forests and ferns. The first reptiles evolve from amphibians, and insects abound.
290–248 million (Permian)	The Earth cools and the great forests disappear. Amphibians decrease in number, and reptiles diversify. Many species disappear in the largest known mass extinction.
248–206 million (Triassic)	The climate returns to being warm again. Conifers and ferns form large forests. The first dinosaurs and ammonites appear.
206–144 million (Jurassic)	In a very moist climate, there are flowering ferns and some conifers. Reptiles predominate, dinosaurs diversify on land, and ammonites reach their height in the seas. The first birds appear.
144–65 million (Cretaceous)	Aquatic reptiles and ammonites reach the height of their importance. On land, ferns give way to willows, maple trees, elms, and flowering plants. At the end of the Cretaceous Period, flying reptiles grow to large sizes. At this time, dinosaurs, swimming and flying reptiles, as well as ammonites and belemnites, become extinct.
65–24 million (Paleogenic)	Different groups of mammals expand and populate the ecosystems left vacant by the disappearance of the dinosaurs. The first ungulates, lemurs, and archaic carnivores appear, and marsupials and insect-eating mammals develop. Flowering plants and pollinating insects continue to evolve and spread widely.
24–2 million (Neogenic)	The climate becomes increasingly colder and drier. The woods begin to shrink, and mammals reach the height of their diversity. The first hominids appear.
2 million–10,000 (Pleistocene)	Various glaciations take place, and many mammal species, including mammoths and saber-tooth tigers, disappear. At the end of this period, *Homo sapiens* appear.
10,000–present (Holocene)	Human beings develop agriculture and technology. They increase in population, while some animal and plant species become extinct as a result of the contamination and occupation of their habitats by humans.

INDEX